THOMAS CRANE PUBLIC LIBRARY
QUINCY MASS
CITY APPROPRIATION

math standards workout

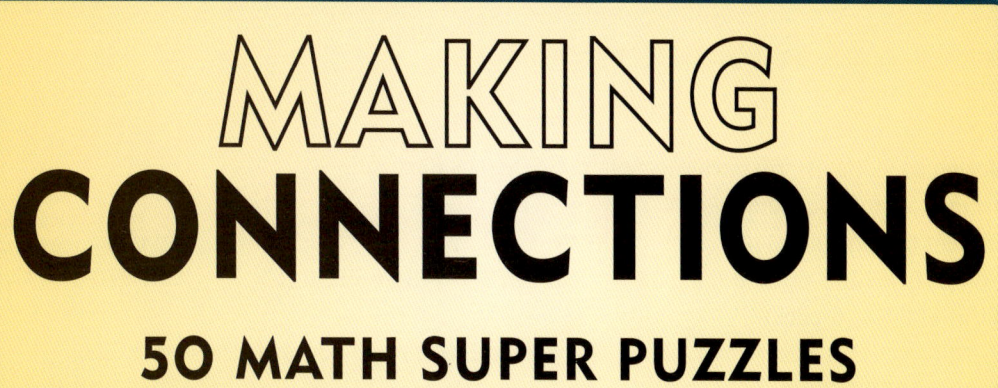

MAKING CONNECTIONS
50 MATH SUPER PUZZLES

By Thomas Canavan

This edition first published in 2012 by The Rosen Publishing Group, Inc.
29 East 21st Street, New York, NY 10010

Copyright © 2012 Arcturus Publishing Limited

All rights reserved. No part of this book may be reproduced in any form without permission in writing from the publisher, except by a reviewer.

Author: Thomas Canavan
Editor: Joe Harris
Design: Jane Hawkins
Cover design: Jane Hawkins

Library of Congress Cataloging-in-Publication Data

Canavan, Thomas, 1956-
Making connections : 50 math super puzzles / Thomas Canavan.
p. cm. — (Math standards workout)
Includes bibliographical references and index.
ISBN 978-1-4488-6674-8 (lib. bdg.) — ISBN 978-1-4488-6681-6 (pbk.) — ISBN 978-1-4488-6687-8 (6-pack)
1. Mathematics—Juvenile literature. 2. Mathematical recreations—Juvenile literature. I. Title.
QA40.5.C355 2003
510—dc23
2011028985

Printed in China
SL002074US

CPSIA Compliance Information: Batch #W12YA. For further information, contact Rosen Publishing, New York, New York, at 1-800-237-9932.

Contents

Introduction	4
Number Crunch	5
Number Path	6
Hexagony	7
Mini Sudoku	8
Tile Twister	9
One to Nine	10
Number Crunch	11
Making Arrangements	12
Total Concentration	13
What's the Number?	14
Number Path	15
Tile Twister	16
Number Crunch	17
One to Nine	18
What's the Number?	19
Hexagony	20
Making Arrangements	21
Tile Twister	22
Number Path	23
What's the Number?	24
Mini Sudoku	25
Circling In	26
Number Crunch	27
Number Path	28
Tile Twister	29
Total Concentration	30
Making Arrangements	31
What's the Number?	32
Number Path	33
Mini Sudoku	34
Circling In	35
Making Arrangements	36
Total Concentration	37
Circling In	38
Word Puzzles	39
Solutions	40
Glossary	46
Further Information	47
Index	48

Introduction

Why do you need this book?
Just as sports such as soccer and basketball include individual skills that you can practice—such as dribbling, passing, and shooting—math problem-solving calls on a range of different skills. Each of the 50 puzzles in *Making Connections* challenges you to find links, patterns, or a new way of looking at a problem to find its solution. These puzzles will help you to build on your real strengths and improve in your weaker areas.

How will this book help you at school?
Making Connections complements the National Council of Teachers of Mathematics (NCTM) framework of Math Standards, providing an engaging enhancement of the curriculum in the following areas:

> *Algebra: Understand Patterns, Relations, and Functions*
> *Connection Standards 3--5*

Why have we chosen these puzzles?
This *Math Standards Workout* title features a range of interesting and absorbing puzzle types, challenging students to master the following skills to arrive at solutions:

- Describe, extend, and make generalizations about geometric and numeric patterns: e.g. Hexagony, What's the Number?

- Represent and analyze patterns and functions: e.g. Mini Sudoku, Tile Twister

- Recognize and use connections among mathematical ideas: e.g. Total Concentration

- Understand how mathematical ideas interconnect and build on one another to produce a coherent whole: e.g. Circling In, Number Path

NOTE TO READERS
If you have borrowed this book from a school or classroom library, please respect other students and DO NOT write your answers in the book. Always write your answers on a separate sheet of paper.

Number Crunch

Starting at the top left with the number provided, work down from one box to another, applying the mathematical instructions to your running total. Write your answers on a separate sheet of paper.

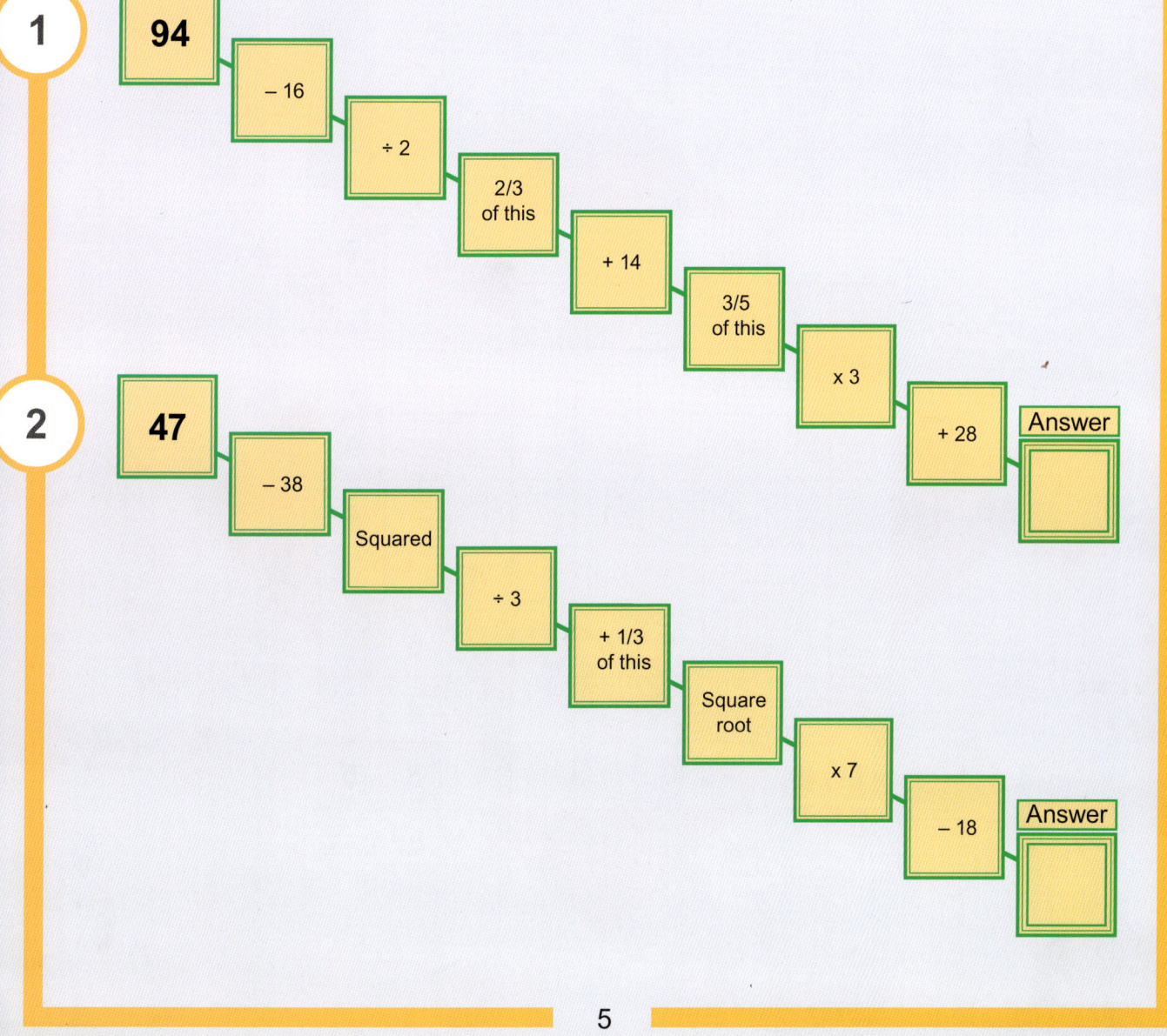

Number Path

Working from one square to another, horizontally or vertically (never diagonally), draw paths to pair up each set of two matching numbers. No path may be shared, and none may enter a square containing a number or part of another path. Write your answers on a separate sheet of paper.

3

Hexagony

Can you place the hexagons into the grid, so that where any hexagon touches another along a straight line, the number in both triangles is the same? No rotation of any hexagon is allowed! Write your answers on a separate sheet of paper.

4

Mini Sudoku

Every row, column, and each of the four smaller boxes of four squares should contain a different number from 1 to 4 inclusive. Some numbers are already in place. Can you complete the grid? Write your answers on a separate sheet of paper.

5

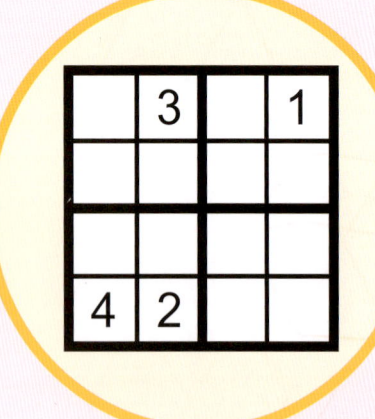

6

	4		
			3
1			
		2	

7

Every row, column, and each of the six smaller boxes of four squares should contain a different number from 1 to 6 inclusive. Some numbers are already in place. Can you complete the grid? Write your answers on a separate sheet of paper.

2	5	6	4		
1					
				1	
		3		6	5
	6				4
5	4		2		

Tile Twister

Place the eight tiles into the puzzle grid so that all adjacent numbers on each tile match up. Tiles may be rotated through 360 degrees, but none may be flipped over. Write your answers on a separate sheet of paper.

8

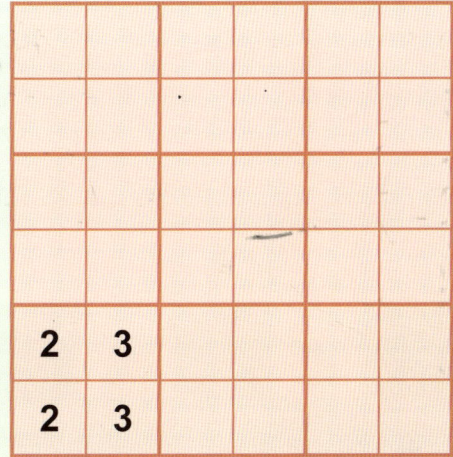

One to Nine

Using the numbers one to nine, complete these six equations (three reading across and three reading downward). Every number is used once only, and one is already in place. Write your answers on a separate sheet of paper.

9

1 2 3 4 5 6 7 8 9

10

Number Crunch

Starting at the top left with the number provided, work down from one box to another, applying the mathematical instructions to your running total. Write your answers on a separate sheet of paper.

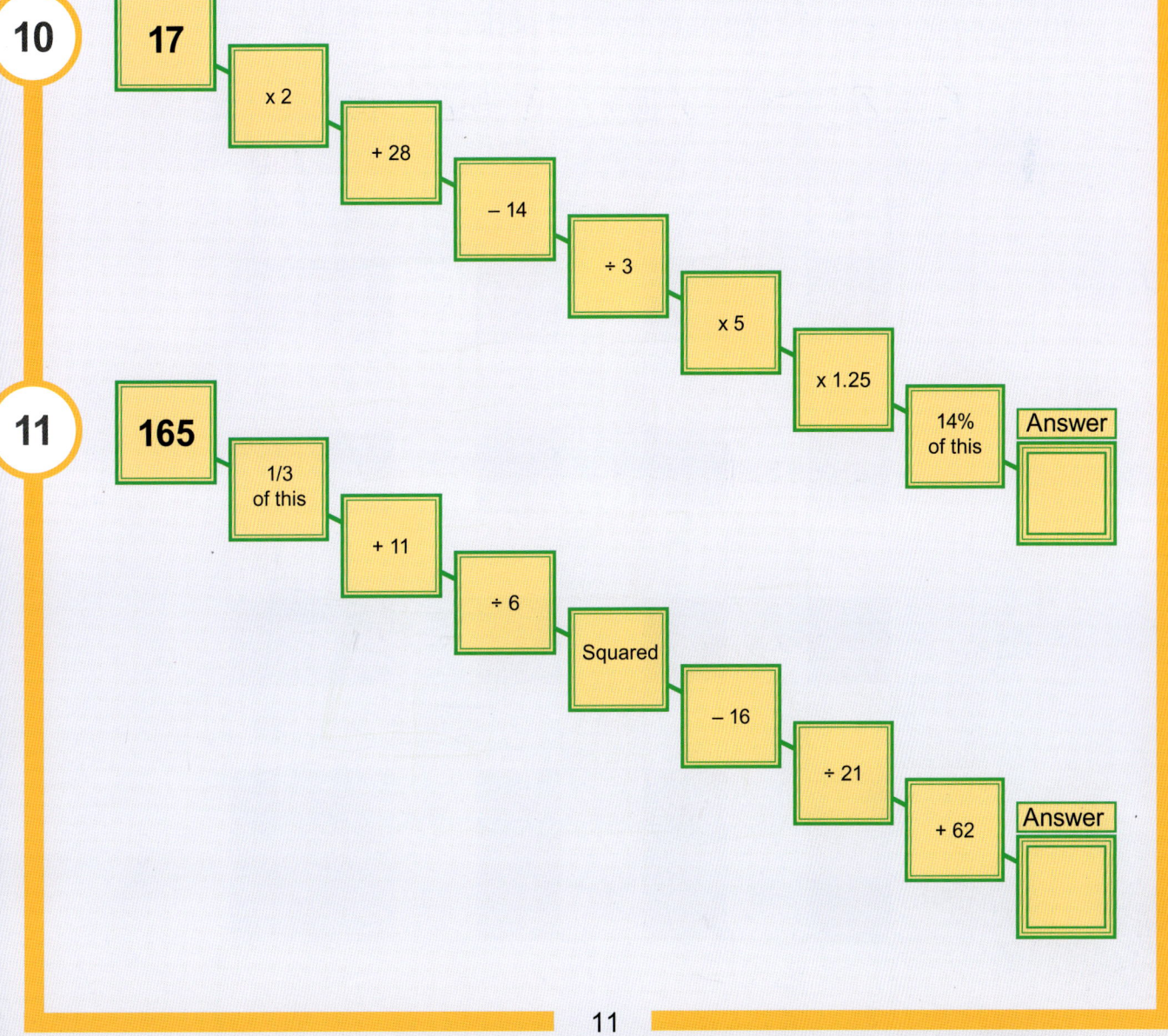

10 17 → × 2 → + 28 → − 14 → ÷ 3 → × 5 → × 1.25 → 14% of this → Answer

11 165 → 1/3 of this → + 11 → ÷ 6 → Squared → − 16 → ÷ 21 → + 62 → Answer

Making Arrangements

Arrange one each of the four numbers below, as well as one each of the symbols x (times), – (minus), and + (plus) in every row and column. You should arrive at the answer at the end of the row or column, making the calculations in the order in which they appear. Some are already in place. Write your answers on a separate sheet of paper.

12

2 5 6 8

Total Concentration

The blank squares below should contain whole numbers between 1 and 30 inclusive, any of which may occur more than once, or not at all. The numbers in every horizontal row add up to the totals on the right, as do the two long diagonal lines extending from corner to corner; those in every vertical column add up to the totals along the bottom. Write your answers on a separate sheet of paper.

108

17	16		25		2	11	82
3	3		13	7	20		67
	29	24	19	10		6	101
15		6		11	26		69
30		21	28	13		17	140
8	22	18	14		27	14	108
	19	9		18	5	4	88

| 108 | 107 | 97 | 110 | 68 | 96 | 69 | 89 |

What's the Number?

In the diagram below, what number should replace the question mark? Write your answer on a separate sheet of paper.

14

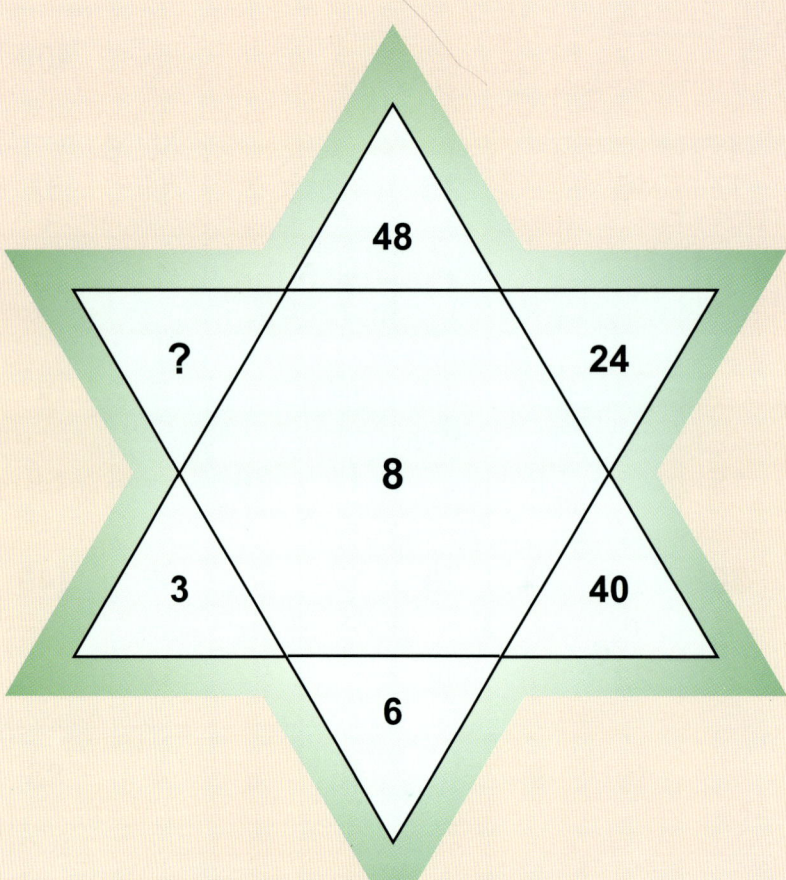

Number Path

Working from one square to another, horizontally or vertically (never diagonally), draw paths to pair up each set of two matching numbers. No path may be shared, and none may enter a square containing a number or part of another path. Write your answers on a separate sheet of paper.

15

Tile Twister

Place the eight tiles into the puzzle grid so that all adjacent numbers on each tile match up. Tiles may be rotated through 360 degrees, but none may be flipped over. Write your answers on a separate sheet of paper.

Number Crunch

Starting at the top left with the number provided, work down from one box to another, applying the mathematical instructions to your running total. Write your answers on a separate sheet of paper.

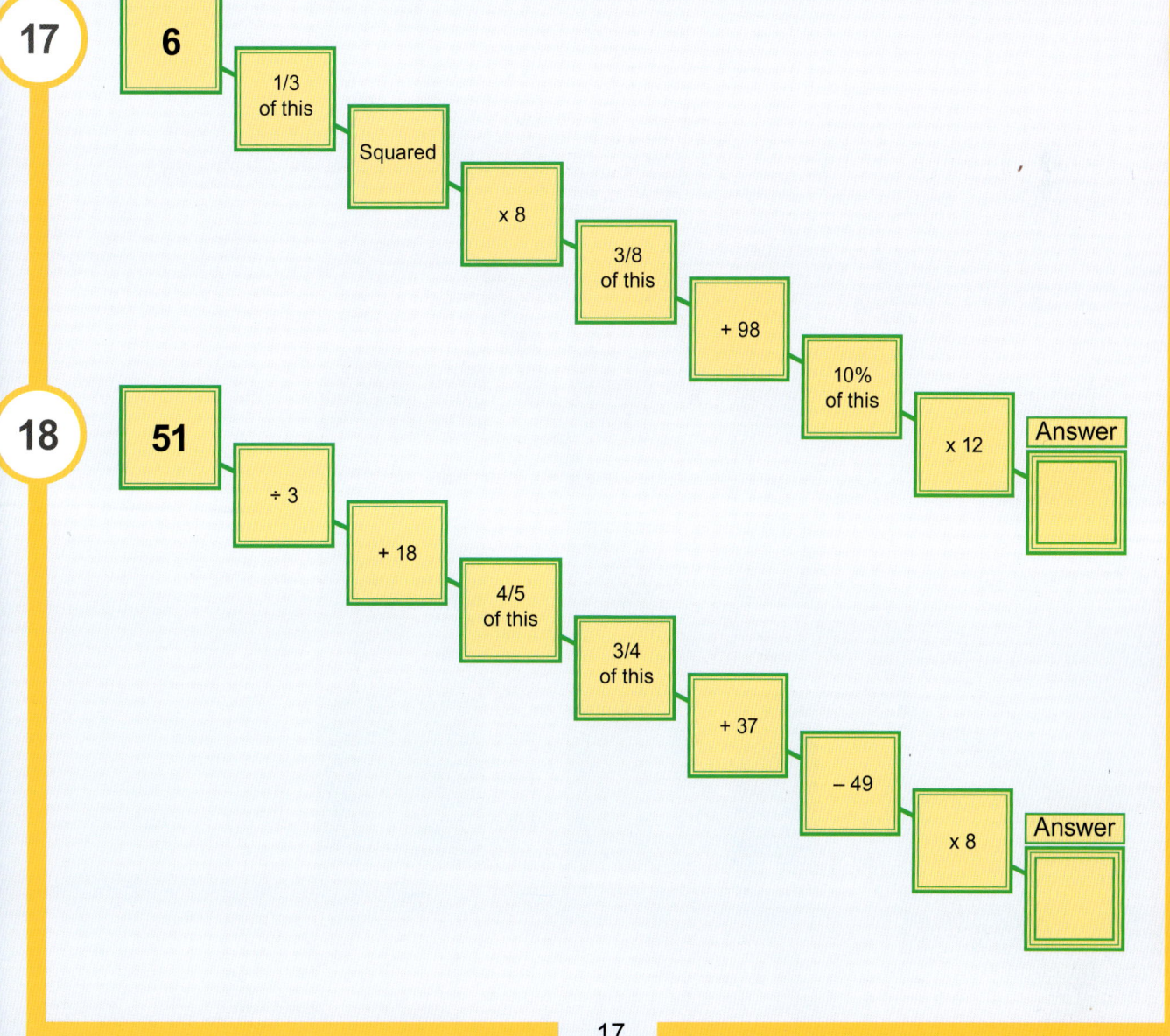

17 6 → 1/3 of this → Squared → × 8 → 3/8 of this → + 98 → 10% of this → × 12 → Answer

18 51 → ÷ 3 → + 18 → 4/5 of this → 3/4 of this → + 37 → − 49 → × 8 → Answer

One to Nine

19

Using the numbers one to nine, complete these six equations (three reading across and three reading downward). Every number is used once only, and one is already in place. Write your answers on a separate sheet of paper.

1 2 3 4 5 6 7 8 9

What's the Number?

In the diagram below, what number should replace the question mark? Write your answer on a separate sheet of paper.

20

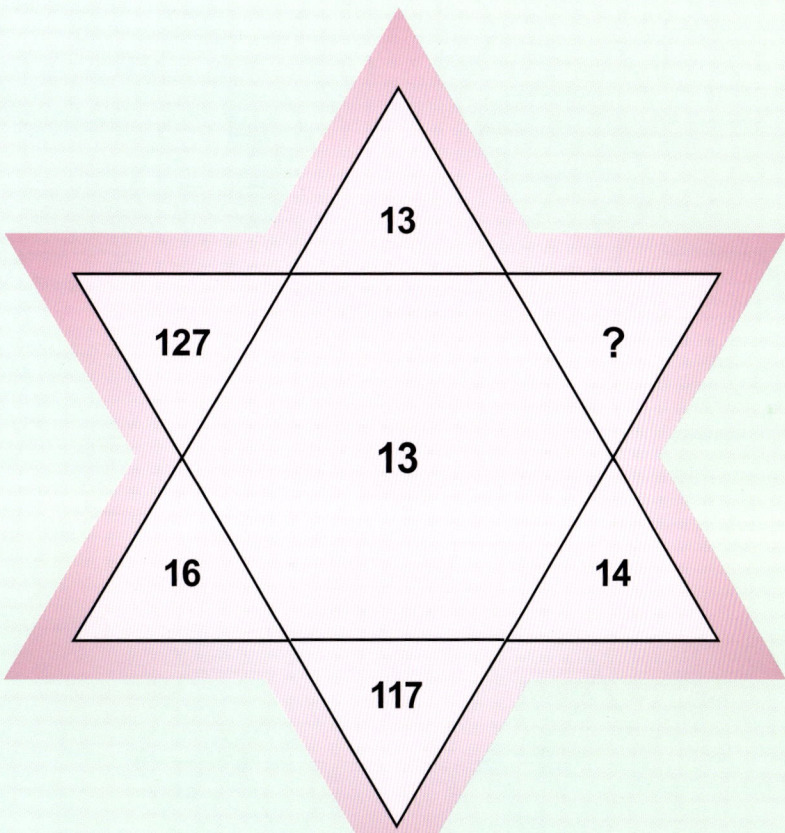

Hexagony

Can you place the hexagons into the grid, so that where any hexagon touches another along a straight line, the number in both triangles is the same? No rotation of any hexagon is allowed! Write your answers on a separate sheet of paper.

21

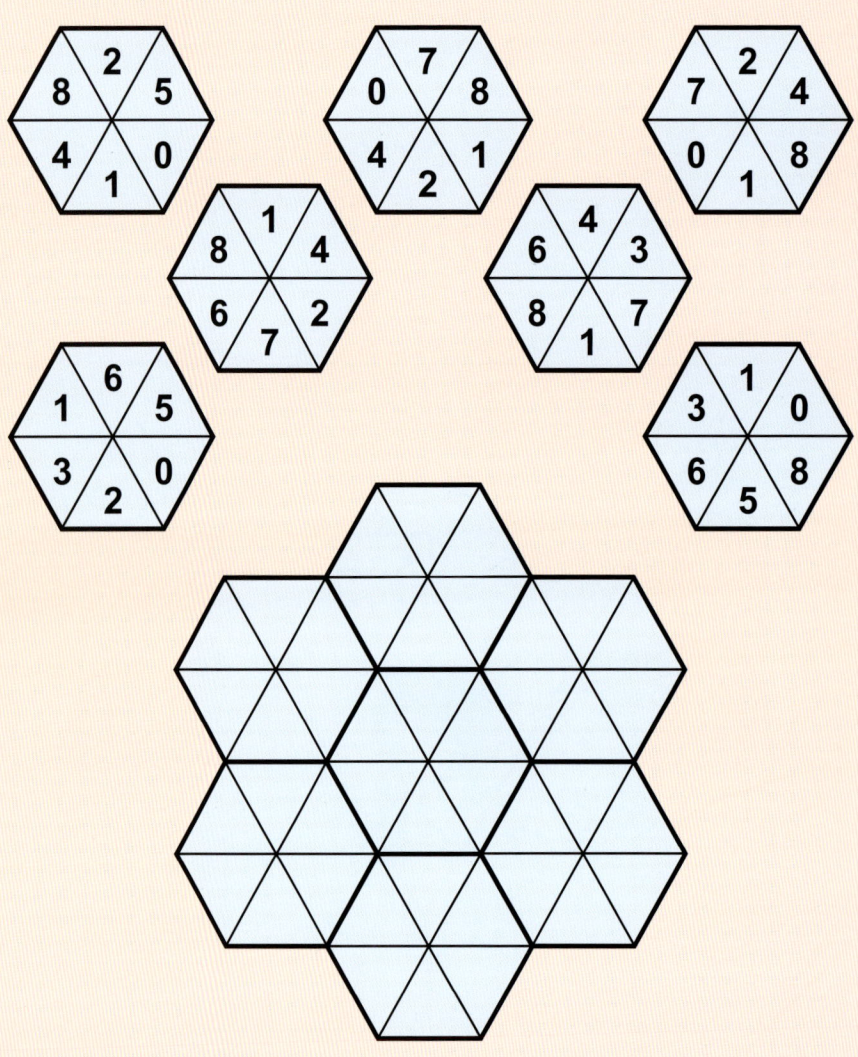

Making Arrangements

Arrange one each of the four numbers below, as well as one each of the symbols x (times), – (minus), and + (plus) in every row and column. You should arrive at the answer at the end of the row or column, making the calculations in the order in which they appear. Some are already in place. Write your answers on a separate sheet of paper.

2 5 7 9

Tile Twister

23

Place the eight tiles into the puzzle grid so that all adjacent numbers on each tile match up. Tiles may be rotated through 360 degrees, but none may be flipped over. Write your answers on a separate sheet of paper.

Number Path

Working from one square to another, horizontally or vertically (never diagonally), draw paths to pair up each set of two matching numbers. No path may be shared, and none may enter a square containing a number or part of another path. Write your answers on a separate sheet of paper.

What's the Number?

In the diagram below, what number should replace the question mark? Write your answer on a separate sheet of paper.

25

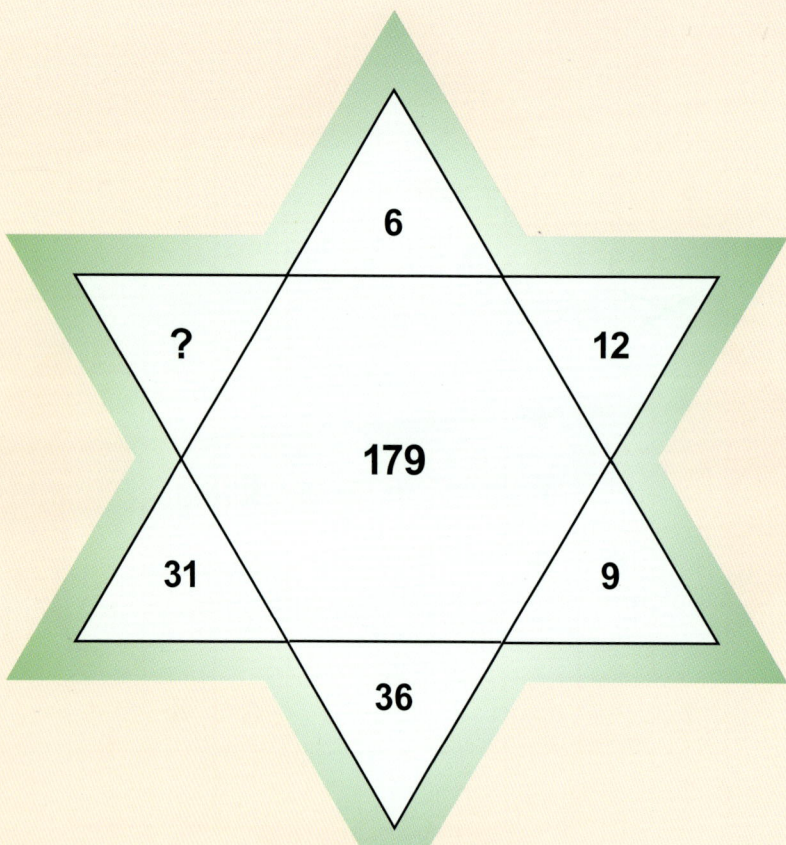

Mini Sudoku

Every row, column, and each of the four smaller boxes of four squares should contain a different number from 1 to 4 inclusive. Some numbers are already in place. Can you complete the grid? Write your answers on a separate sheet of paper.

26

	2		
		3	
1			4

27

	4		2
3			
			3
		4	

Every row, column, and each of the six smaller boxes of four squares should contain a different number from 1 to 6 inclusive. Some numbers are already in place. Can you complete the grid? Write your answers on a separate sheet of paper.

28

	1			5	
					3
3		2			
4	5				2
1	2	4		6	
5	3		1		

Circling In

The three empty circles should contain the symbols +, −, and x in some order, to make a series that leads to the number in the middle. Each symbol must be used once and calculations are made in a clockwise direction. Write your answers on a separate sheet of paper.

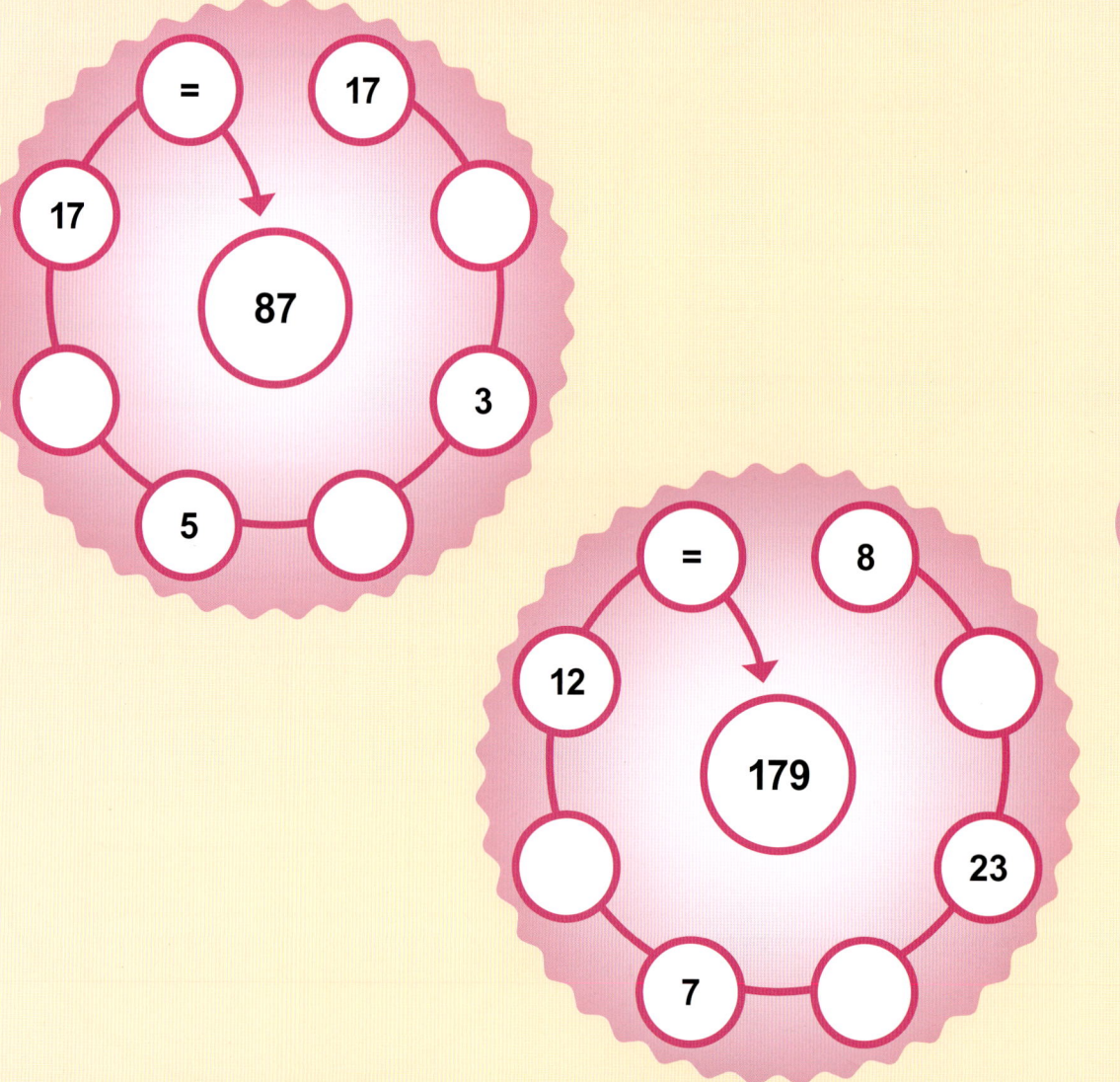

Number Crunch

Starting at the top left with the number provided, work down from one box to another, applying the mathematical instructions to your running total. Write your answers on a separate sheet of paper.

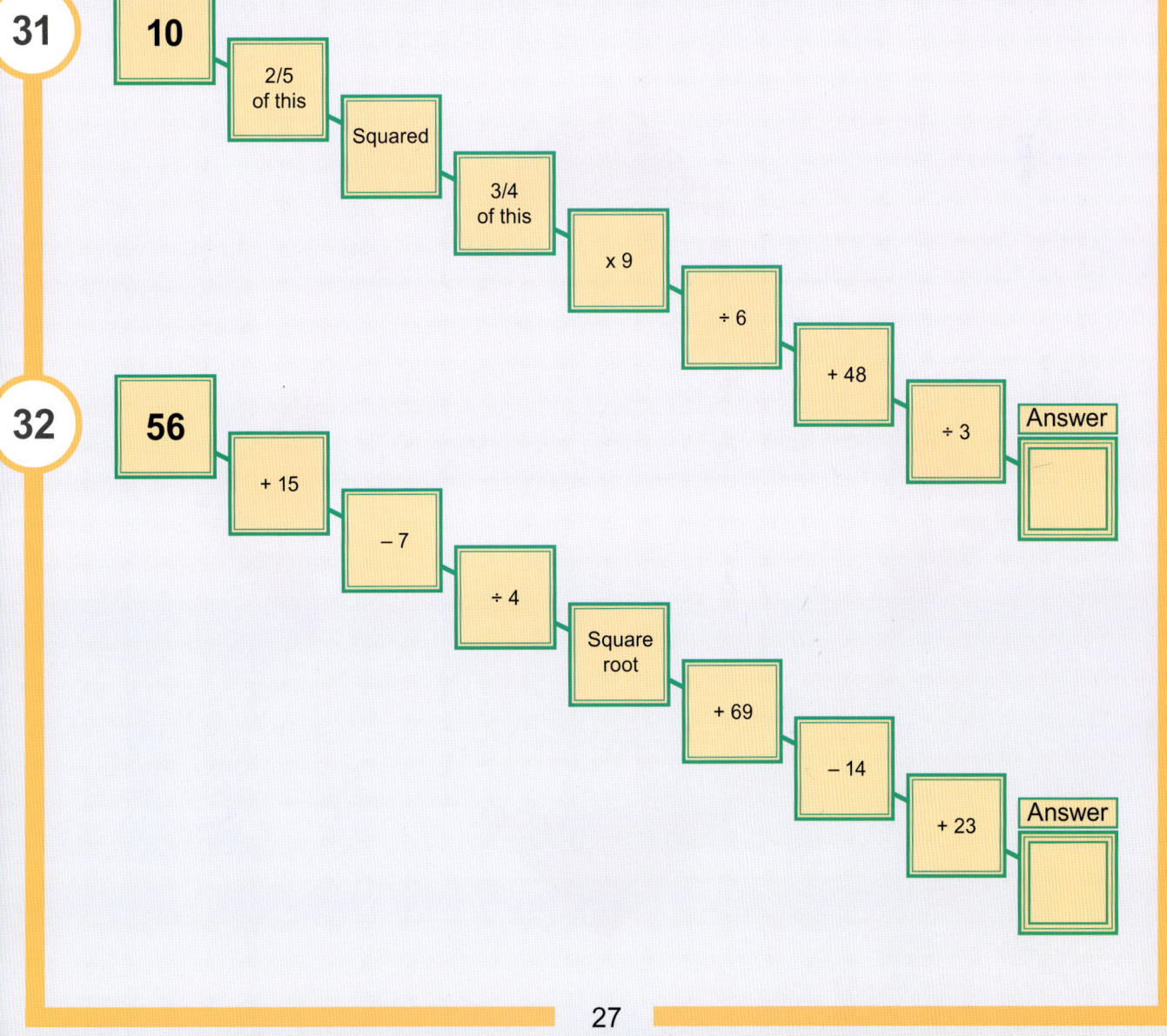

Number Path

Working from one square to another, horizontally or vertically (never diagonally), draw paths to pair up each set of two matching numbers. No path may be shared, and none may enter a square containing a number or part of another path. Write your answers on a separate sheet of paper.

33

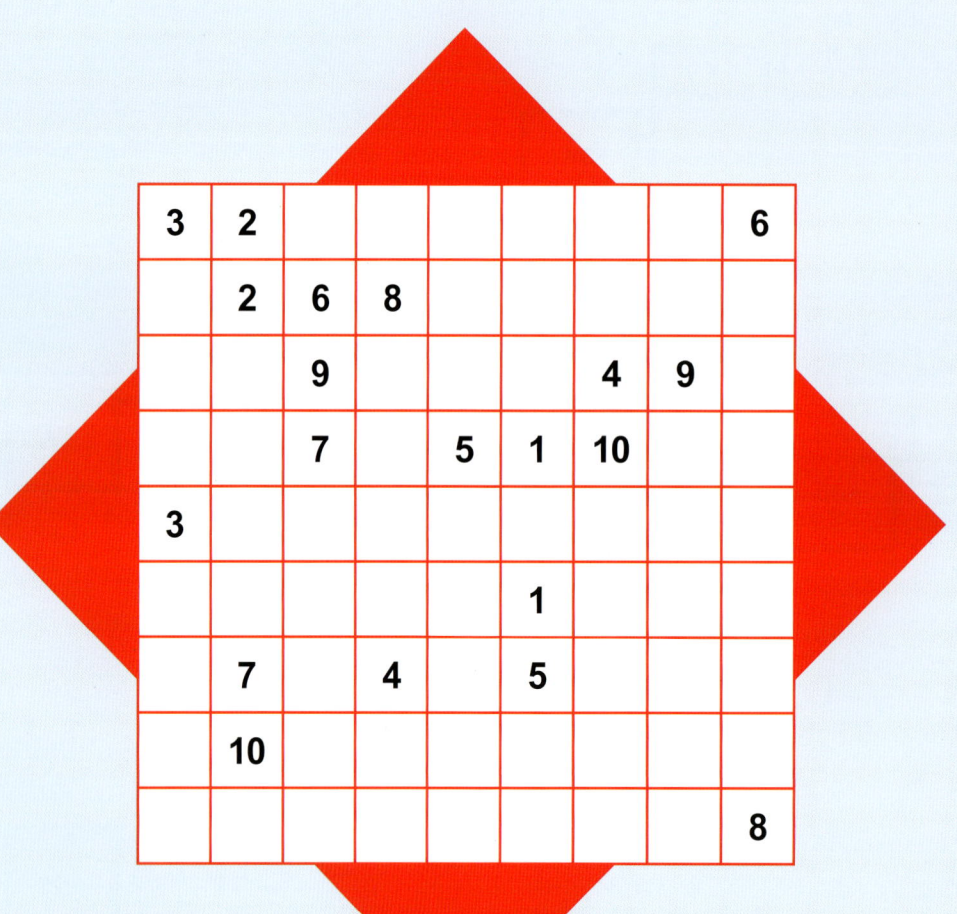

Tile Twister

Place the eight tiles into the puzzle grid so that all adjacent numbers on each tile match up. Tiles may be rotated through 360 degrees, but none may be flipped over. Write your answers on a separate sheet of paper.

34

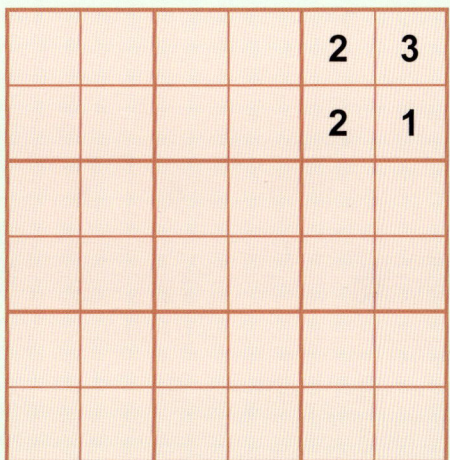

Total Concentration

35

The blank squares below should contain whole numbers between 1 and 30 inclusive, any of which may occur more than once, or not at all. The numbers in every horizontal row add up to the totals on the right, as do the two long diagonal lines extending from corner to corner; those in every vertical column add up to the totals along the bottom. Write your answers on a separate sheet of paper.

							123
21	27		1	4	6		108
20	9	16		22	5		104
2		8	15	5		8	90
10	2	23	22		1		103
	6	27	13		24	18	137
	25	14	3		11	29	92
20	4			12	24	23	142
104	99	147	109	88	97	132	115

30

Making Arrangements

Arrange one each of the four numbers below, as well as one each of the symbols x (times), – (minus), and + (plus) in every row and column. You should arrive at the answer at the end of the row or column, making the calculations in the order in which they appear. Some are already in place. Write your answers on a separate sheet of paper.

36

1 3 8 9

What's the Number?

In the diagram below, what number should replace the question mark? Write your answer on a separate sheet of paper.

37

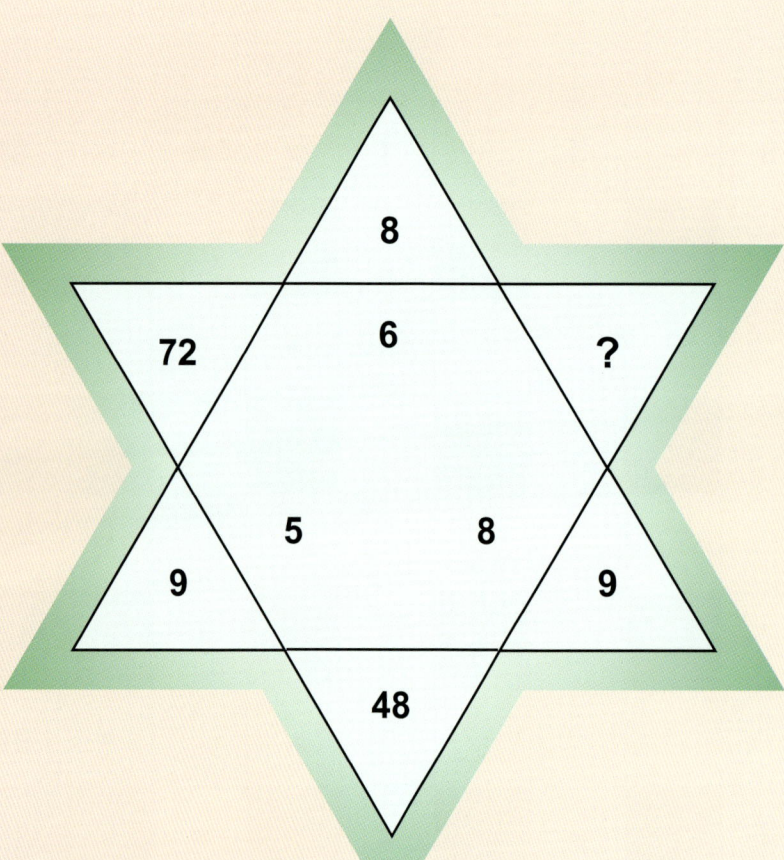

32

Number Path

Working from one square to another, horizontally or vertically (never diagonally), draw paths to pair up each set of two matching numbers. No path may be shared, and none may enter a square containing a number or part of another path. Write your answers on a separate sheet of paper.

38

9	10	5						6
				7				
					5	8	10	
			6		3			
					3			
		8						
			9	4	2			
4					2	7	1	1

33

Mini Sudoku

Every row, column, and each of the four smaller boxes of four squares should contain a different number from 1 to 4 inclusive. Some numbers are already in place. Can you complete the grid? Write your answers on a separate sheet of paper.

39

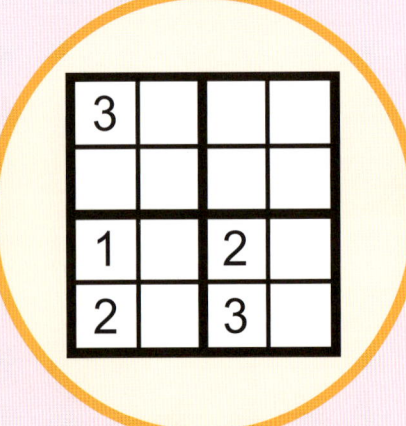

3			
1		2	
2		3	

40

		2	
3	2		
			1
4			

41

Every row, column, and each of the six smaller boxes of four squares should contain a different number from 1 to 6 inclusive. Some numbers are already in place. Can you complete the grid? Write your answers on a separate sheet of paper.

3				2	
2			1	4	
	5	2			6
6	4			1	
					1
			3		

Circling In

The three empty circles should contain the symbols +, −, and x in some order, to make a series that leads to the number in the middle. Each symbol must be used once and calculations are made in a clockwise direction. Write your answers on a separate sheet of paper.

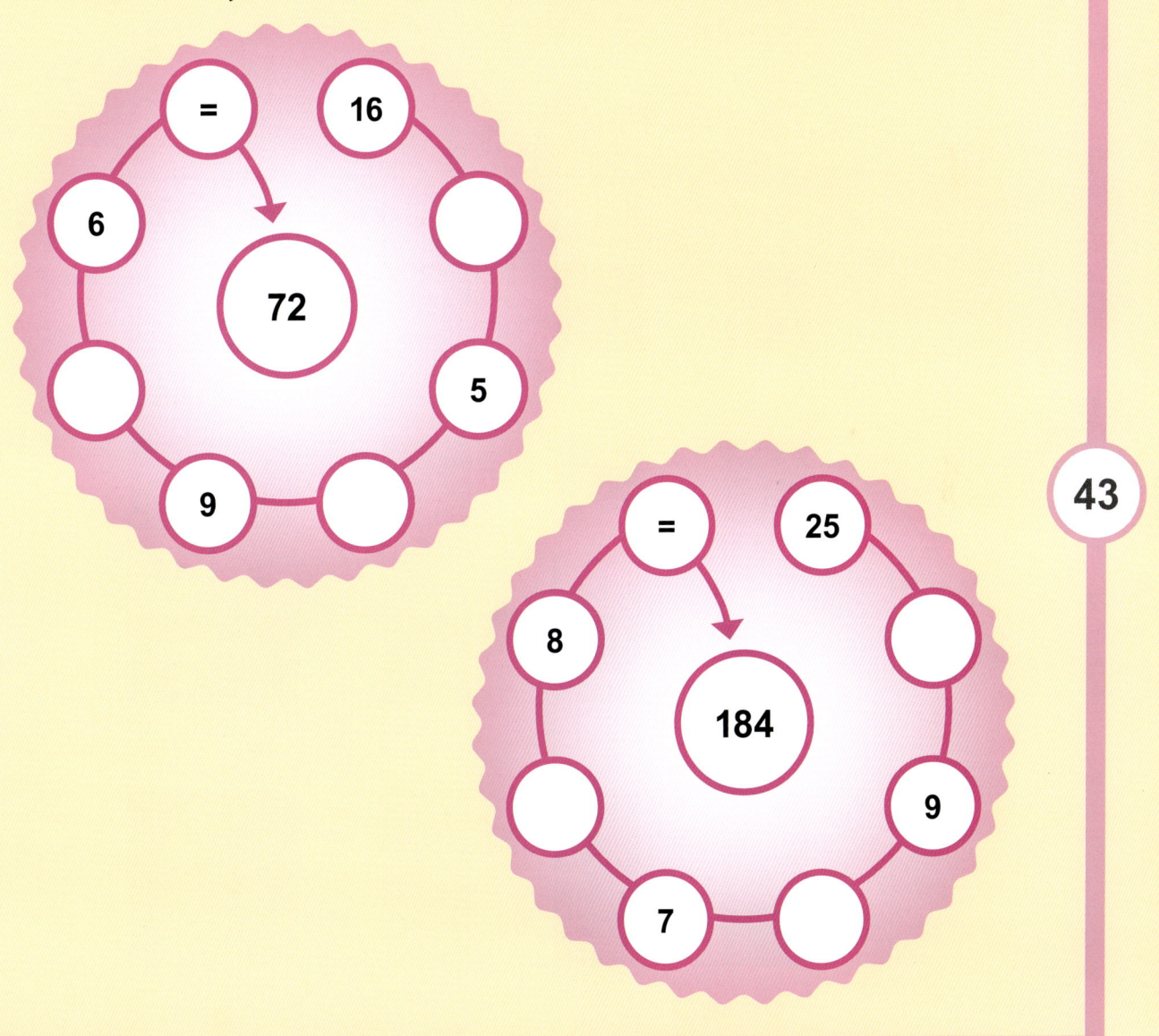

42

Circle 1: 16, 5, 9, 6 with = and 72 in the middle

43

Circle 2: 25, 9, 7, 8 with = and 184 in the middle

Making Arrangements

44

Arrange one each of the four numbers below, as well as one each of the symbols x (times), – (minus), and + (plus) in every row and column. You should arrive at the answer at the end of the row or column, making the calculations in the order in which they appear. Some are already in place. Write your answers on a separate sheet of paper.

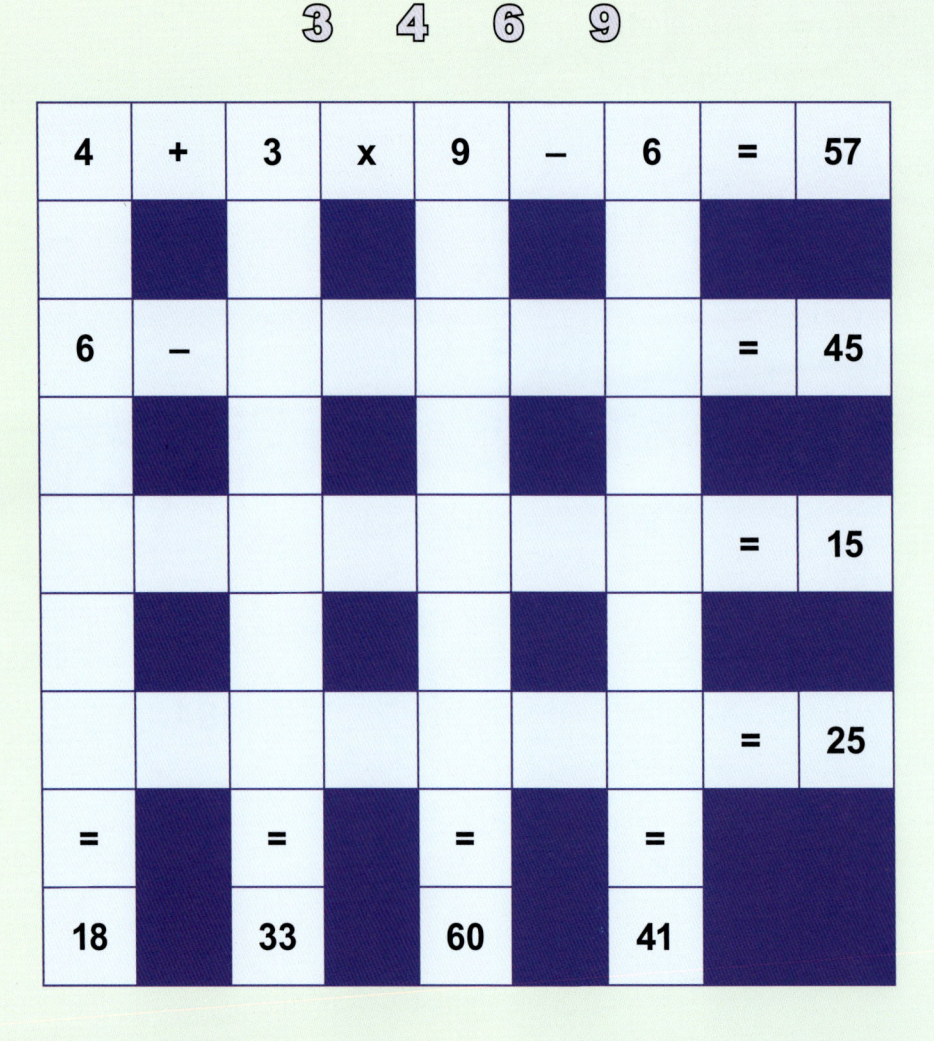

3 4 6 9

4	+	3	x	9	–	6	=	57
6	–						=	45
							=	15
							=	25
=		=		=		=		
18		33		60		41		

Total Concentration

The blank squares below should contain whole numbers between 1 and 30 inclusive, any of which may occur more than once, or not at all. The numbers in every horizontal row add up to the totals on the right, as do the two long diagonal lines extending from corner to corner; those in every vertical column add up to the totals along the bottom. Write your answers on a separate sheet of paper.

45

85

17		10		14	3	20	100
	3		15	8	14	19	107
29	16		2		21	4	92
4	11		9		27	13	79
30		5	22	19		26	119
18	23	7		18	12		96
	11	17	6		16		100
119	97	83	79	103	98	114	91

37

Circling In

The three empty circles should contain the symbols +, –, and x in some order, to make a series that leads to the number in the middle. Each symbol must be used once and calculations are made in a clockwise direction. Write your answers on a separate sheet of paper.

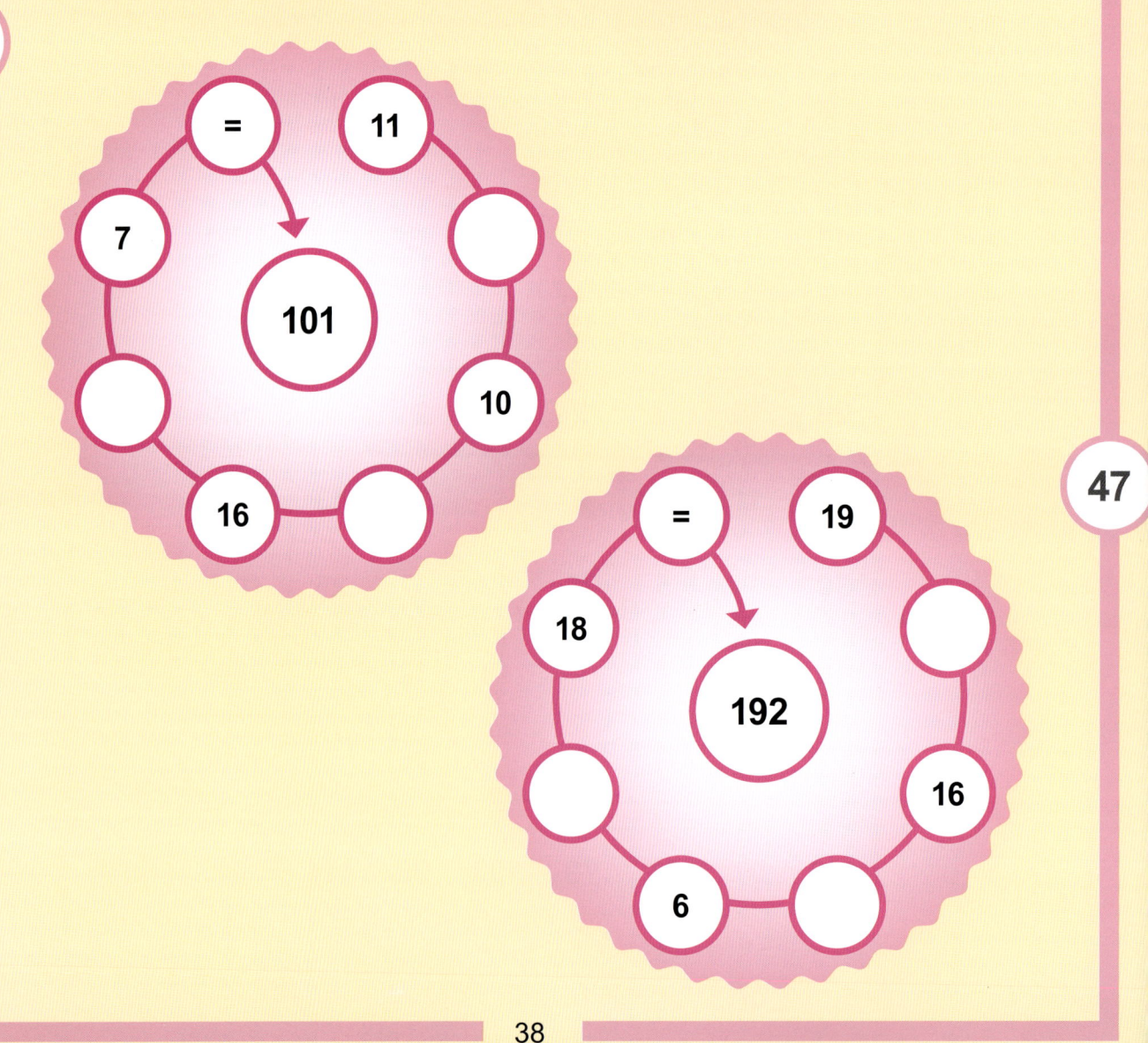

46

47

Word Puzzles

48 **Are You Getting Warmer?**
If 0 degrees Celsius is the same as 32 degrees Fahrenheit, and 100 degrees Celsius is the same as 212 degrees Fahrenheit, what is the Fahrenheit equivalent of 10 degrees Celsius?

49 **The Right Note**
On your trip to Spain, you noticed that one Euro was worth $1.40 and that you could change your dollars into Euros at banks and post offices. You saw a used Spanish guitar at a flea market, but it cost 70 Euros. You have $100—will you be able to buy the guitar?

50 **Relative Values**
If your father has three sisters, each of whom has two sons, and your mother has two brothers, each of whom has three daughters, then how many first cousins do you have?

Solutions

1

94 − 16 = 78, 78 ÷ 2 = 39, 39 ÷ 3 x 2 = 26, 26 + 14 = 40, 40 ÷ 5 x 3 = 24, 24 x 3 = 72, 72 + 28 = 100

2

47 − 38 = 9, 9^2 = 81, 81 ÷ 3 = 27, 27 + 9 = 36, square root of 36 = 6, 6 x 7 = 42, 42 − 18 = 24

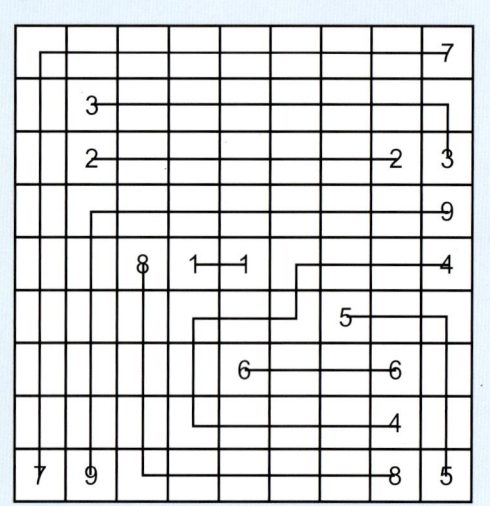

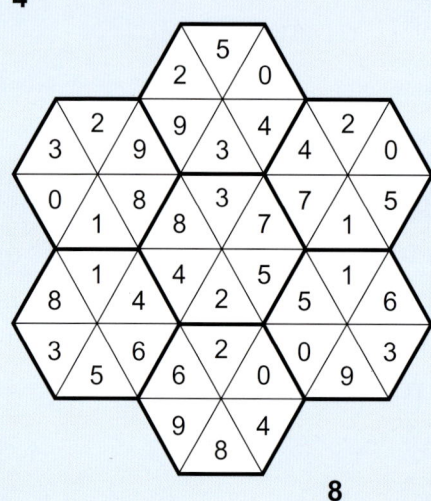

3

4

5

2	3	4	1
1	4	3	2
3	1	2	4
4	2	1	3

6

4	3	1	2
2	1	3	4
1	4	2	3
3	2	4	1

7

2	5	6	4	3	1
1	3	4	5	2	6
6	2	5	3	1	4
4	1	3	6	5	2
3	6	2	1	4	5
5	4	1	2	6	3

8

4	4	4	2	2	4
3	2	2	1	1	3
3	2	2	1	1	3
2	3	3	2	2	1
2	3	3	2	2	1
2	3	3	4	4	3

9

9	−	2	+	7	=	14
−		+		−		
4	+	8	−	3	=	9
x		x		+		
1	x	6	+	5	=	11
=		=		=		
5		60		9		

10

17 x 2 = 34, 34 + 28 = 62, 62 − 14 = 48, 48 ÷ 3 = 16, 16 x 5 = 80, 80 x 1.25 = 100, 14% of 100 = 14

11

165 ÷ 3 = 55, 55 + 11 = 66, 66 ÷ 6 = 11, 11^2 = 121, 121 − 16 = 105, 105 ÷ 21 = 5, 5 + 62 = 67

Solutions

12

5	x	8	+	2	−	6	=	36
+		−		x		+		
6	−	2	x	5	+	8	=	28
x		+		−		x		
8	−	5	+	6	x	2	=	18
−		x		+		−		
2	+	6	x	8	−	5	=	59
=		=		=		=		
86		66		12		23		

13

17	16	7	25	4	2	11	82
3	3	12	13	7	20	9	67
12	29	24	19	10	1	6	101
15	2	6	1	11	26	8	69
30	16	21	28	13	15	17	140
8	22	18	14	5	27	14	108
23	19	9	10	18	5	4	88
108	107	97	110	68	96	69	89

108

14

5 – Each single digit number in the outer point of the star is multiplied by the central number to reach the total in the opposite point of the star.

15

16

3	2	2	3	3	3
4	4	4	1	1	2
4	4	4	1	1	2
3	1	1	1	1	2
3	1	1	1	1	2
2	2	2	4	4	3

17

6 ÷ 3 = 2, 2^2 = 4, 4 x 8 = 32, 32 ÷ 8 x 3 = 12, 12 + 98 = 110, 10% of 110 = 11, 11 x 12 = 132

18

51 ÷ 3 = 17, 17 + 18 = 35, 35 ÷ 5 x 4 = 28, 28 ÷ 4 x 3 = 21, 21 + 37 = 58, 58 − 49 = 9, 9 x 8 = 72

Solutions

19

4	x	3	−	7	=	5
+		x		+		
1	x	6	+	9	=	15
x		+		−		
8	−	5	x	2	=	6
=		=		=		
40		23		14		

20

147 – Each two-digit number is multiplied by 10 and the central number is deducted from this sum to give the number in the opposite point of the star.

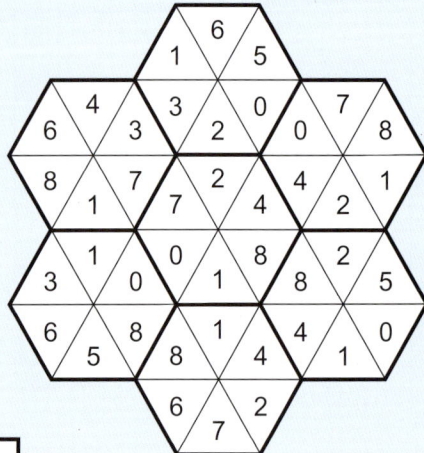

21

22

5	+	7	−	2	x	9	=	90
−		+		x		−		
2	x	5	+	9	−	7	=	12
x		x		−		+		
9	−	2	x	7	+	5	=	54
+		−		+		x		
7	+	9	x	5	−	2	=	78
=		=		=		=		
34		15		16		14		

23

4	3	3	1	1	4
4	2	2	2	2	4
4	2	2	2	2	4
3	3	3	4	4	1
3	3	3	4	4	1
1	2	2	3	3	4

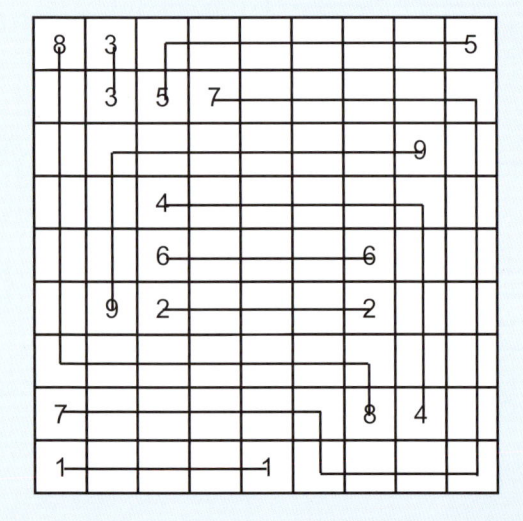

24

42

Solutions

25

186 – Working clockwise from the top, multiply the first number by two, then deduct three, then multiply by four, then deduct five, then multiply by six (the question mark is replaced by 186), then deduct seven to reach the number in the middle.

26

3	2	4	1
4	1	3	2
2	4	1	3
1	3	2	4

27

1	4	3	2
3	2	1	4
4	1	2	3
2	3	4	1

28

2	1	3	4	5	6
6	4	5	2	1	3
3	6	2	5	4	1
4	5	1	6	3	2
1	2	4	3	6	5
5	3	6	1	2	4

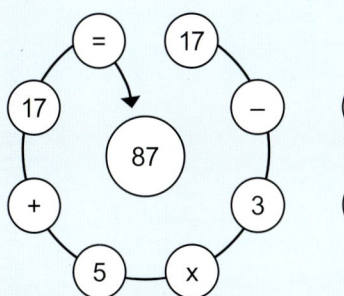

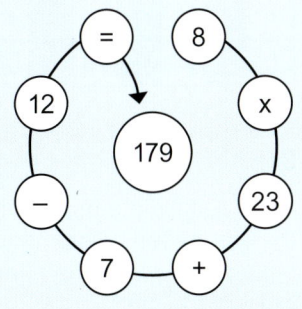

29

17 =, 17, +, 5, x, 3, −, center: 87

30

12 =, 8, x, 23, +, 7, −, center: 179

31

10 ÷ 5 x 2 = 4, 4^2 = 16, 16 ÷ 4 x 3 = 12, 12 x 9 = 108, 108 ÷ 6 = 18, 18 + 48 = 66, 66 ÷ 3 = 22

32

56 + 15 = 71, 71 − 7 = 64, 64 ÷ 4 = 16, square root of 16 = 4, 4 + 69 = 73, 73 − 14 = 59, 59 + 23 = 82

33

(grid solution)

34

2	3	3	2	2	3
1	1	1	2	2	1
1	1	1	2	2	1
2	4	4	3	3	3
2	4	4	3	3	3
1	4	4	1	1	4

43

Solutions

35

							123
21	27	30	1	4	6	19	108
20	9	16	25	22	5	7	104
2	26	8	15	5	26	8	90
10	2	23	22	17	1	28	103
28	6	27	13	21	24	18	137
3	25	14	3	7	11	29	92
20	4	29	30	12	24	23	142
104	99	147	109	88	97	132	115

36

9	−	3	+	1	×	8	=	56
+		×		+		−		
1	×	9	−	8	+	3	=	4
×		−		×		+		
3	+	8	×	9	−	1	=	98
−		+		−		×		
8	−	1	+	3	×	9	=	90
=		=		=		=		
22		20		78		54		

37

45 – Multiply the number in the outer point of the star by the adjacent number in the central hexagon to reach the number in the opposite point of the star.

38

39

3	1	4	2
4	2	1	3
1	3	2	4
2	4	3	1

40

1	4	2	3
3	2	1	4
2	3	4	1
4	1	3	2

41

3	1	4	6	2	5
2	6	5	1	4	3
1	5	2	4	3	6
6	4	3	5	1	2
4	3	6	2	5	1
5	2	1	3	6	4

44

Solutions

42

6 = 16 + 5 − 9 × → 72

43

8 = 25 − 9 + 7 × → 184

44

4	+	3	×	9	−	6	=	57
×		+		−		+		
6	−	4	+	3	×	9	=	45
−		×		+		×		
9	−	6	×	4	+	3	=	15
+		−		×		−		
3	×	9	−	6	+	4	=	25
=		=		=		=		
18		33		60		41		

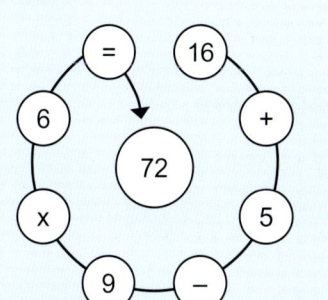

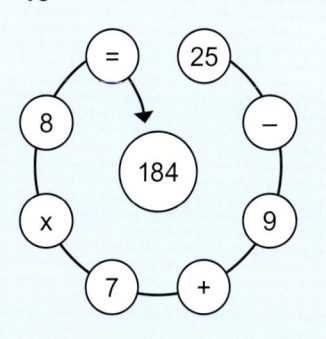

45

17	21	10	15	14	3	20	100
20	3	28	15	8	14	19	107
29	16	7	2	13	21	4	92
4	11	9	9	6	27	13	79
30	12	5	22	19	5	26	119
18	23	7	10	18	12	8	96
1	11	17	6	25	16	24	100
119	97	83	79	103	98	114	91

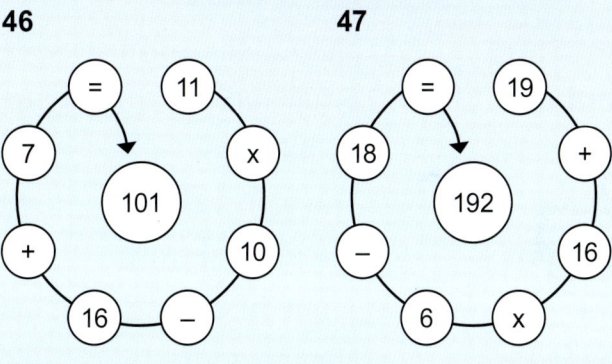

85

46

7 = 11 × 10 − 16 + → 101

47

18 = 19 + 16 × 6 − → 192

48

50 degrees

49

Yes, because the $100 is worth 71.43 Euros.

50

12 first cousins

45

Glossary

adjacent	Close to or—more commonly—next to.
calculation	The use of math to find a solution.
column	A line of objects that goes straight up and down.
concentration	Thinking very hard and examining every possibility.
diagonal	Moving in a slanted direction, halfway between straight across and straight down.
grid	A display of crisscrossed lines.
hexagon	A six-sided object.
horizontal	A direction that is straight across.
inclusive	Including both ends of a series (two to five inclusive means 2, 3, 4, and 5).
matching	Exactly the same as.
mini	Small (an informal word).
occur	To happen.
rotate	Travel in a circular motion.
rotation	Circular motion.
row	A line of objects that goes straight across.
shared	Having something the same as something else.
square root	A number that, if multiplied by itself, produces the original number (3 is the square root of 9; 4 is the square root of 16.)
squared	When a number is multiplied by itself (3 squared = 3 x 3 = 9).
vertical	A direction that is straight up and down.
whole number	A number that has no decimals (4 is a whole number; 4.3 is not a whole number).

Further Information

For More Information

Consortium for Mathematics (COMAP)
175 Middlesex Turnpike, Bedford, MA 01730
(800) 772-6627 http://www.comap.com/index.html
COMAP is a nonprofit organization whose mission is to improve mathematics education for students of all ages. It works with teachers, students, and business people to create learning environments where mathematics is used to investigate and model real issues in our world.

MATHCOUNTS Foundation
1420 King Street, Alexandria, VA 22314
(703) 299-9006 https://mathcounts.org/sslpage.aspx
MATHCOUNTS is a national enrichment, club, and competition program that promotes middle school mathematics achievement. To secure America's global competitiveness, MATHCOUNTS inspires excellence, confidence, and curiosity in U.S. middle school students through fun and challenging math programs.

National Council of Teachers of Mathematics (NCTM)
906 Association Drive, Reston, VA 20191-1502
(703) 620-9840 http://www.nctm.org
The NCTM is a public voice of mathematics education supporting teachers to ensure equitable mathematics learning of the highest quality for all students through vision, leadership, professional development and research.

Web Sites

Due to the changing nature of Internet links, Rosen Publishing has developed an online list of Web sites related to the subject of this book. This site is updated regularly. Please use this link to access this list:
> http://www.rosenlinks.com/msw/conn

Further Reading

Abramson, Marcie F. *Painless Math Word Problems.* New York, NY: Barron's Educational Series, 2010.
Fisher, Richard W. *Mastering Essential Math Skills: 20 Minutes a Day to Success (Book Two: Middle Grades/High School).* Los Gatos, CA: Math Essentials, 2007.

Lewis, Barry. *Help Your Kids With Math: A Visual Problem Solver for Kids and Parents.* New York, NY: DK Publishing, 2010.

Overholt, James, and Laurie Kincheloe. *Math Wise! Over 100 Hands-On Activities That Promote Real Math Understanding.* New York, NY : Jossey-Bass, 2010.

Index

adjacent 9, 16, 22, 29

calculations 12, 21, 26, 31, 35, 36, 38
Celsius 39
Circling In 26, 35, 38
clockwise 26, 35, 38
column 8, 12, 13, 21, 25, 30, 31, 34, 36, 37
cousins 39

degrees 9, 16, 22, 29, 39
diagonal 13, 30, 37
diagonally 6, 15, 23, 28, 33
diagram 14, 19, 24, 32
dollars 39

equations 10, 18
Euros 39

Fahrenheit 39
flipped 9, 16, 22, 29

grid 7, 8, 9, 16, 20, 22, 25, 29, 34

hexagons 7, 20
Hexagony 7, 20
horizontal 13, 30, 37
horizontally 6, 15, 23, 28, 33

inclusive 8, 25, 34

Making Arrangements 12, 21, 31, 36
matching 6, 15, 23, 28, 33
Mini Sudoku 8, 25, 34

Number Crunch 5, 11, 17, 27
Number Path 6, 15, 23, 28, 33

One to Nine 10, 18

paths 6, 15, 23, 28, 33

rotated 9, 16, 22, 29
rotation 7, 20
row 8, 12, 13, 21, 25, 30, 31, 34, 36, 37

squared 5, 11, 17, 27
square root 5, 27
star 41, 44
sudoku 8, 25, 34
symbols 12, 21, 26, 31, 35, 36, 38

Tile Twister 9, 16, 22, 29
Total Concentration 13, 30, 37

vertical 13, 30, 37
vertically 6, 15, 23, 28, 33

What's the Number? 14, 19, 24, 32
whole number 13, 30, 37
Word Puzzles 39